Angelo Lourenco Guimaraes

Environmental Impacts of Stone Extraction for Construction

Angelo Lourenco Guimaraes

Environmental Impacts of Stone Extraction for Construction

Case studies

ScienciaScripts

Imprint

Any brand names and product names mentioned in this book are subject to trademark, brand or patent protection and are trademarks or registered trademarks of their respective holders. The use of brand names, product names, common names, trade names, product descriptions etc. even without a particular marking in this work is in no way to be construed to mean that such names may be regarded as unrestricted in respect of trademark and brand protection legislation and could thus be used by anyone.

Cover image: www.ingimage.com

This book is a translation from the original published under ISBN 978-620-2-17422-0.

Publisher:
Sciencia Scripts
is a trademark of
Dodo Books Indian Ocean Ltd. and OmniScriptum S.R.L publishing group

120 High Road, East Finchley, London, N2 9ED, United Kingdom
Str. Armeneasca 28/1, office 1, Chisinau MD-2012, Republic of Moldova, Europe
Printed at: see last page
ISBN: 978-620-7-30375-5

Index :

INTRODUCTION

The environment and sustainable development are issues that have gradually gained an important position and are now the subject of study in various sciences. Environmental impacts have created a growing interest in recent times due to the importance given to environmental management and the preservation of natural resources. This makes it necessary to deepen our knowledge of the negative effects produced on the environment by human activities, with the aim of preventing or reducing their impact.

Continuous socio-economic expansion, reflected in the accelerated urbanization resulting from the development of the industrial and service sectors, population growth, among others, has increased the demand for natural resources and, in particular, for mineral resources that are directly used in construction, fostering an extractive activity that transforms the environment and induces its degradation.

It is in this context that this paper arises, with a view to addressing the environmental impacts resulting from stone quarrying in Mafuiane and the intensity of its impacts on the environment and the social life of the population. It should be noted that the approach focuses on the negative impacts of stone quarrying.

In terms of the structure of the work, apart from the introductory aspects where we give a brief presentation of the work, it is made up of three (3) chapters.

In the first chapter, we try to take a theoretical approach to the problem at hand, i.e. a literature review.

The second chapter is reserved for the generalities of the study area, specifically the physical-geographical and socio-economic aspects.

The third chapter deals with the problem of stone quarrying in Mafuiane and its environmental impacts. This chapter analyzes and discusses the results of the fieldwork by interpreting the data from the surveys, always establishing a relationship with the legislation applicable to this activity. Finally, there are the conclusions and recommendations. The work also includes a bibliography, appendices and annexes.

Theme and justification

The title of this paper is *"Environmental problems in the Sulbrita quarry in the town of Mafuiane"*. It is part of the Education and Sustainable Development (ESD) research line at the Geography Department of the Pedagogical University (UP).

More specifically, it was the result of a study visit to the District of Namaacha in 2006 as part of an academic excursion led by Prof. Dr. Gustavo Dgedge, then lecturer in Geomorphology, and the following year, I visited the same region in the same perspective in the company of Prof. Dr. Mueca and Mestre Sabil. I was able to see that stone extraction in the Mafuiane region jeopardizes the quality of the environment, since this activity is carried out without observing the minimum environmental protection rules suggested by the mining legislation in force in the country.

It also arises from the desire to contribute to the knowledge of environmental problems that arise as a consequence of the unsustainable extraction of natural resources and to contribute to the search for possible solutions that could minimize environmental damage.

This fact, combined with the reflection provided by the experiences and research carried out during the course in various curricular subjects, in addition to the desire to contribute scientific knowledge, gave rise to the idea of making available to academics and why not to society in general and to mining operators in that region some information that is considered important for the process of environmental preservation and conservation, which is a current bet with a view to promoting Sustainable Development (SD).

Problematization

The mining industry supplies materials for a wide variety of constructions, such as roads, railways, airports, ports and other civil engineering works, as well as raw materials for the manufacture of cement and other derivatives. These materials are obtained by exploiting quarries and sand pits.

Although mineral extraction is important for socio-economic development in this region, it is also responsible for negative environmental impacts that are often irreversible because it affects the mining site and its surroundings, causing various impacts on the environment. However, the activity of exploiting natural resources causes changes to the environment, the so-called environmental impacts (EI), because the nature of the processes involved in mining, in particular the extraction of stone, cause damage to the environment and to people, and can even affect existing buildings around the quarries.

However, there is a need to reduce the volume of materials extracted in quarries in order to protect the environment and conserve the quality and quantity of the resources around the quarries, a situation that is not happening in the study area, or if it does happen, there is not much demand from the competent authorities.

So, based on the above assumptions, the following starting question arises:
- ***What is the environmental impact of Sulbrita's quarry in Mafuiane, Namaacha District?***

Subject of study

The environmental impacts caused by the Sulbrita quarry in the town of Mafuiane, in the district of Namaacha, are the subject of this study.

Hypothesis

- If the Sulbrita quarry complied with the recommendations of the Simplified Environmental Impact Assessment Study and the rules established in the mining license in force in the country during the extraction of the stone, the impacts on the environment would be minimized.

Objectives of the work

General:

- Assessing the environmental impacts caused by the Sulbrita quarry in Mafuiane, Namaacha District.

Specific:

- Describe the physical-geographical and socio-economic aspects of the town of Mafuiane in the district of Namaacha;
- Assessing the scale of the environmental impacts caused by the Sulbrita quarry in Mafuiane, Namaacha District;
- To assess the environmental impacts caused by the Sulbrita quarry in the context of stone extraction in the locality of Mafuiane, Namaacha District;
- Propose measures to solve the environmental problems caused by the Sulubrita quarry in the locality of Mafuiane, Namaacha District.

Working methodology

The research took place between September 2009 and September 2010 and was divided into four phases: Preparation of the research project, literature review, data collection, analysis and interpretation of the results. To this end, a qualitative investigation was carried out using the following methods: Bibliographical Review, Direct and Indirect Observation, Cartographic Method, Statistical Mathematical Method, Semi-structured Interview Technique and Analysis and Synthesis Method.

1. Literature review

With this method, we wanted to understand everything that had been written about the problem under analysis. So this operation consisted of collecting geological data from the National Geology Directorate (DNG), the Mogambique Agricultural Research Institute (IIAM), the Ministry for the Coordination of Environmental Action (MICOA), the Provincial Directorate for the Coordination of Environmental Action, the Geology Department Library (DG) at Eduardo Mondlane University (UEM), the Brazao Mazula library and the UP-Sede library.

This information was obtained from books, especially AFONSO et al. (1998), GUALE (1999), FRANCISCO & PASSELA (2007), MAE (2005), MUCHANGOS (1999), NHATUGÉS (2003), PASSELA (2006, 2007), PIRES (1995), SZABO (1985), VILANCULOS & SERNO (1993).

2. Direct observation method

It took place between March and June 2010, and consisted of a trip to the District of Namaacha, and in particular to the town of Mafuiane, where the geographical, socio-economic and environmental conditions were observed on the ground, as well as the exploitation of mineral resources, the main environmental impacts caused by stone extraction, such as the emission of dust, fumes and noise pollution, and the taking of photos documenting the facts that were uncovered.

3. Cartographic method

We consulted the Soil Chart of Maputo and Gaza Province on a scale of 1:5000, the Geological Chart on a scale of 1:250,000, sheet 2532, the Geological Chart of the People's Republic of Mozambique on a scale of 1: 1,000,000, the Topography Chart on a scale of 1:50,000 and the map of the administrative division of Namaacha on a scale of 1:400,000 and 1:570,000.

4. Statistical mathematical method

We used this method to draw up tables and graphs, as well as to process and interpret the figures in the work.

5. Indirect observation method

It enabled data to be collected on: geographical location, physical and socio-economic information on the study area, using observation tools such as questionnaires and interviews.

6. Semi-structured interview technique

We interviewed the head of the Mafuiane locality, the neighborhood secretary, the head of the health center and workers from the Sulbrita quarry. The planned sample was 80 people, but it was only possible to work with 50 people, including 1 head of Mafuiane locality, 1 secretary of the neighborhood, 1 head of the Mafuiane health center, 2 shopkeepers, 39 peasants living on the left bank of the EN2 in the area adjacent to the Sulbrita quarry, specifically the Bacabaca neighborhood, 6 Sulbrita quarry workers selected at random during the fieldwork in the exploration area (see Table 1 and Figure 1).

Table 1: Sample distribution

Interviewees	Occupation
1	Head of Mafuiane Locality
1	Secretary of the Bacabaca Neighborhood
1	Head of the Mafuiane Health Center
2	Traders
6	Sulbrita quarry workers
39	Peasants

Source: author, 2010

Fig.1: Distribution diagram of the interview sample by sex, age and occupation

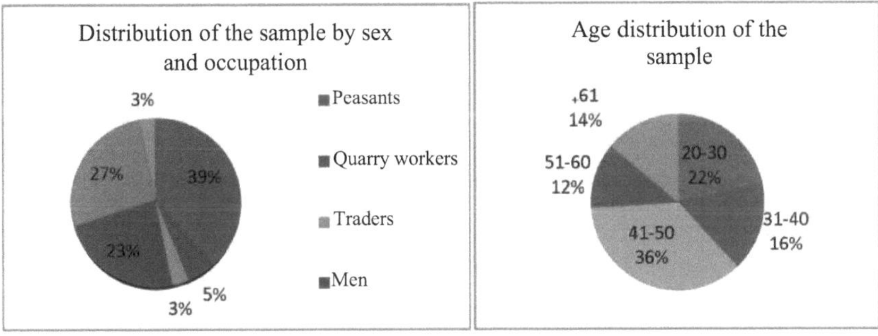

Source: author, 2010

For the local chief, the neighborhood secretary, the head of the health center and the quarry workers, the means of communication was Portuguese, and for the peasants, given the difficulties in speaking Portuguese, we used the local language (Ronga and Xangana), for which we had the support of the neighborhood secretary to translate the questions we asked.

In order to ensure that all the answers given during the interview were obtained, we recorded the interview using a "*Nokia 6300*" cell phone, and it is with this phone that we took the photographs that are included in the work.

7. Method of analysis and synthesis

We used this method to analyze and interpret the data collected in the field to create the final content of the work.

CHAPTER 1

GENERAL ASPECTS OF THE EXPLOITATION OF MINING RESOURCES IN MOZAMBIQUE

1. Literature review and theoretical framework

Centuries ago, man had a relatively moderate impact on nature, which did not substantially jeopardize the ecological balance or deprive future generations of the natural resources they needed to sustain the human species.

Meanwhile, most of the Mogambican population depends on the exploitation of natural resources for their subsistence and income generation. It is in this context that the action plan for the reduction of absolute poverty (2006-2009) PARPA II, in view of this reality, recognizes that the achievement of the objectives depends profoundly on the way in which natural resources are managed and conserved, and on the relationship between their use and exploitation and the benefit for the poor. However, NOTICE (2006:26) pointed out that:

> *"As humanity increases its capacity to intervene in nature to satisfy its growing needs and desires, tensions and conflicts arise over the use of space and resources. From the same perspective, the same author points out that the global demand for natural resources stems from an economic formation based on large-scale production and consumption. The logic associated with this formation, which governs the process of exploiting nature today, is responsible for the destruction of natural resources (...)".*

Therefore, the exploitation of natural resources is a very old practice. For several centuries, the main natural resources, namely minerals, forests and soil, were used to produce goods that were modest at the time, but which constituted the needs of the time.

With the Industrial Revolution (IR) in the 2nd^a half of the 18th century, the exploitation of natural resources gained greater impetus as it mechanized agriculture and equipped man with tools and new techniques and technologies for exploiting the various natural resources.

[11]According to MICOA (1995:78), natural resources "are any part of the natural environment such as air, water, soil, forest, fauna and minerals".

Thus, it can be seen that the exploitation of natural resources has evolved with the evolution of humanity itself, involving not only intellectual groups, but society as a whole.

From a historical point of view, *"the process of mining in the District of Namaacha, particularly in the Locality of Mafuiane, goes back a long way, to around the 1960s, when bentonite and other minerals were extracted using shovels and picks"* (NGOMANE, 2010, cp.).[2]

However, with the evolution of technology, these tools have gradually been replaced by pneumatic excavators and wheel loaders, which are then transported by tractors and dump trucks, accelerating the impact on the environment, particularly pollution[3] of the environment. According to SALOMÁO (2006:19), however, the environment is understood to be:

> *"the environment in which humans and other living beings interact with each other and with the environment itself, and includes air, light, land, water, ecosystems, biodiversity, ecological relationships, all organic and inorganic matter, socio-cultural and economic conditions that affect the life of communities".*

In other words, the interaction of all the natural, artificial and cultural elements that provide for the balanced development of life in all its forms. Thus, we can consider the existence of the artificial environment (urban space); the cultural environment (historical, artistic, archaeological, landscape and tourist heritage); and the natural or physical environment (constituted by the interaction of living beings with their environment).

Environmental pollution covers a range of aspects from air, water and soil contamination to landscape disfigurement and erosion, among others. Despite the dependence between industrialization, production, pollution and population growth, environmental contamination by toxic substances is not a recent problem.

This is because, since the dawn of mankind, toxic products and variable waste have been formed by man's action and carried by rivers, wind and rainwater, proving to be toxic to the environment, or at least a nuisance to it. Therefore, man has always been responsible for environmental pollution in such a variety of ways that a simple list of individual factors is impossible.

It was in this context that the United Nations (UN), twenty years after the Stockholm Conference (1972), met from 3 to 14 June 1992 to address environmental issues, holding the Conference on Environment and Sustainable Development[4] (ADS), which became known as the Rio Conference. This conference definitively established international environmental law by determining that environmental impact assessment as a national instrument must be carried out in relation to certain activities that may have a significant adverse impact on the environment and are dependent on a decision by the competent national authorities. It is in this perspective that the Earth has been recognized as our home, of an interdependent and integral nature, and it is proclaimed that human beings are at the center of concerns about Sustainable Development, having the right to a healthy and

---

[2] Personal communication.

3GARRIDO & COSTA (1996:134) define pollution as *"the modification of a natural environment by the introduction of harmful substances or in excess, which hinder the life and evolution of the environment itself. This can be chemical, atmospheric, noise or visual pollution".*

[4] According to the Environment Act Law N.º 20/97 of 1997, Sustainable Development *"is development based on environmental management that satisfies the needs of the present generation without compromising the balance of the environment and the possibility of future generations also satisfying their needs".*

productive life in harmony with nature[5] .

There are several mountain ranges in the world with different elevations, a wide variety of shapes, climates and specific combinations of ecosystems, from the equator to the polar regions. Mountains are one of the most fragile areas on the planet and, despite the efforts of governments and various sectors of society, they are being permanently degraded.

Its resources, especially water and biodiversity, are fundamental to the quality of life of a large part of the population, as well as being an important center for culture and recreation. Worldwide, 10% of inhabitants live in this ecosystem and 40% depend on it directly. However, current trends in globalization, the growth of the construction industry and urbanization are endangering mountain communities and their resources, as they are the target of much exploitation, particularly in the search for building stone, resulting in major environmental impacts.

In Mozambique, there are also several mountains of great ecological and economic importance. For the case under study, in the region of Namaacha in Maputo Province, there is the Libombos Volcanic Chain (CVL), which is also the target of several mining operations by legal and illegal operators, which in a way jeopardize the quality of the local environment.

In November 1998, the (UN) General Assembly declared 2002 as the International Year of Mountains (IYM)[6], with the aim of creating awareness of the global importance of mountain ranges. This decision offered a unique opportunity to reinforce the process started at the (UN) Conference on SDS held in Rio in 1992, to raise public awareness and to ensure adequate political, institutional and financial commitment to anchor the action of Sustainable Mountain Development.

The main result of Rio 92 was a global action program on the environment and development, "Agenda 21".[6][7]

The inclusion of "Chapter 13[8][9] " in the Agenda, entitled "**Sustainable development of** mountains9", placed them on an equal footing with climate change, deforestation in the tropics and other issues of great importance in the global debate on the environment and development. In 1995, the Mountain Forum was founded, a worldwide network of organizations with the aim of disseminating information

---

[5] Principle I of the Rio 92 Conference.

[6] Cf. the article entitled International Year of Mountains available on the website:
http://www.montanhasbrasil.org.br/ano.htm accessed on April 16, 2010.

[7] Agenda 21 was one of the main results of the Rio-92 conference, which took place in Rio de Janeiro, Brazil, in 1992. It is a document that established that each country should commit itself to reflecting, globally and locally, on how governments, companies, non-governmental organizations and all sectors of society could cooperate in the study of solutions to socio-environmental problems.

[8] This chapter recommends, among other things, that states should draw up national legislation on civil liability and compensation for victims of pollution and other environmental damage.

[9] The bold is not original. Sustainable mountain development aims to ensure the present and future well-being of mountain communities by promoting conservation and sustainable development in these areas;
Increasing awareness and knowledge of the mountain ecosystem, its dynamics, the way it functions and its decisive importance in providing certain strategic goods and services that are essential for the well-being of people in the highlands and lowlands, both urban and rural, especially for water supply and food security, promoting and defending the cultural heritage of mountain communities/societies, among others.

on the importance of preserving the environment of mountains and their communities.

We therefore agree that in order to achieve these objectives, there is a need for an exchange of information, sensitization and awareness-raising, education, information, dissemination, documentation of best practices and the formulation of advice based on case studies that have been successful in this field, and the promotion, formulation and specific legislation of a policy on mountains, so that the results of (AIM) go beyond 2002.

From this perspective, the success of AIM requires environmental education, which is a process of recognizing values and clarifying concepts, with the aim of developing skills and changing attitudes towards the environment in order to understand and appreciate the interrelationships between human beings, their cultures and their biophysical environments.

Environmental education tries to make everyone aware that human beings are part of the environment. It tries to overcome the anthropocentric vision, which has made man feel that he is always the center of everything, forgetting the importance of nature, of which he is an integral part.

PASSELA (2006) published the Report of the Environmental Impact Study commissioned by Pedreira do Tamega, which includes the positive and negative impacts resulting from the extraction of stone. In 2007, the same author published the Final Report of the Simplified Environmental Study commissioned by the Riolitos Quarry (Monte Libovane). In the same year, he carried out the environmental impact study requested by the Rugunate quarry.

In 2007, FRANCISCO & PASSELA published the final report of the simplified environmental study requested by Pedreira da Sulbrita, Lda.

From the environmental activity carried out on Mount Secuane, FRANCISCO & PASSELA (2007:37) concluded that:

> *"The most significant positive environmental impacts are associated with the stone extraction activity. These impacts include the generation of dust, noise, surface water pollution, soil pollution, with effects on public health, in particular the workers assigned to the processing plant."*

These impacts include the supply of building stone for various infrastructures underway in the south of the country, creating jobs for local communities. However, the negative impacts are of moderate significance, of medium and high intensity, and are localized to the fact that the area is already altered by mining activity.

CHAPTER 2

PHYSICAL-GEOGRAPHICAL AND SOCIO-ECONOMIC GENERALITIES OF THE TOWN OF MAFUIANE

2.1 Geographical and astronomical location

The town of Mafuiane, formerly known as the communal village of Mafuiane, has existed since 1981 and was founded by a total of 56 families from the Pequenos Libombos reservoir.

Mafuiane is a locality belonging to the Administrative Post of Namaacha and is located 40 kilometers (km) from Maputo City, on the left bank of National Road No. 2 (EN2) which connects Maputo City and the District of Namaacha and on the left bank of the Umbelúzi River, with the following geographical boundaries:

- To the north it is bordered by the town of Kulula;

- To the south it is bordered by the Pequenos Libombos reservoir;

- The east is bordered by the district of Boane,

- To the west it is bordered by Mount Impaputo and the town of Mafavuca 1.

Astronomically, the town of Mafuiane is located between parallels 26° 00' 47" and 26° 03' 21" South Latitude and meridians 32° 13' 11" and 32° 16' 00" East Longitude (see fig.2)[10] .

---

[10] Figure

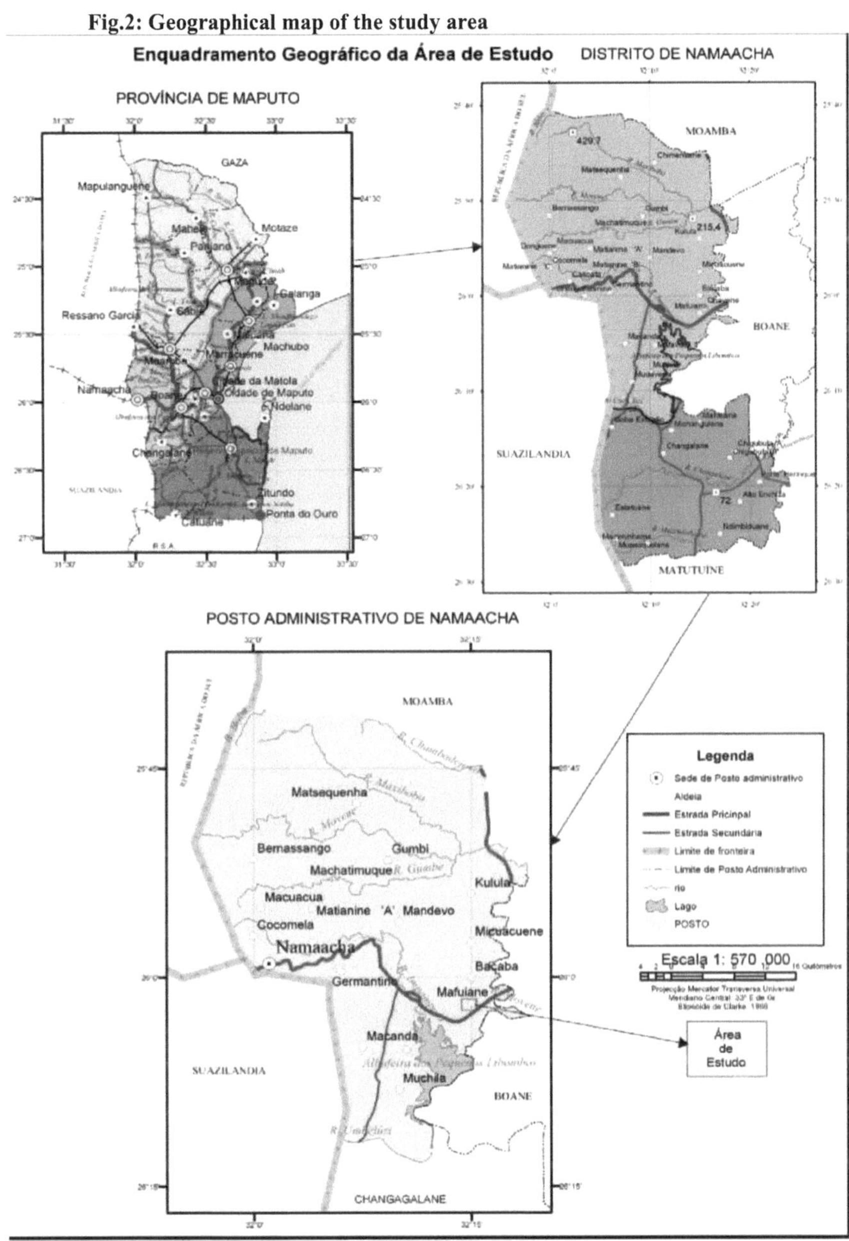

Fonte: CENACARTA 2010

2.1.1 Access to Sulbrita Quarry

Pedreira da Sulbrita Lda. is located in Mafuiane District of Namaacha, Maputo Province, on the right-hand side of the EN2, and its activity is carried out on Monte Secuane. The quarry area covers 160 hectares corresponding to 8 cadastral units whose geographical coordinates are shown in the following table (Tabelal).

The photo below is an aerial view of the area occupied by the Sulbrita quarry (Fig.3). Access to the quarry is via the EN2 highway, which also serves as an outlet for the stone to the markets in the industrial areas of Beleluane, the cities of Matola and Maputo, as well as to the southern region of the country in general.

Table 2: Geographical coordinates of Mount Secuane

Vertex	Latitude (S)	Longitude (E)
1	26°00'30"	32°16'45"
2	26°00'30"	32°17'15"
3	26°01'15"	32°17'15"
4	26°01'15"	32°16'30"
5	26°00'458"	32°16'30"
6	26°00'45"	32°16'45"

Source: FRANCISCO & PASSELA (2007:9)

Fig.3: Satellite image illustrating the surface occupied by the Sulbrita quarry

Source: Google Earth provided on 21/05/10

2.2 Physical-geographical aspects

2.2.1 Weather

When describing geographical and natural features, climate and meteorological phenomena play an important role in understanding them. According to the Koppen classification, the predominant climate throughout most of the District of Namaacha is humid tropical (AW).

As a permanently geographical factor in the climate in the District of Namaacha, we must distinguish the altitude, because according to (BOLEO:1950), "*due to the elevation there is a slight decrease in the average annual temperature from 21° C to 20° C and a sharp increase in rainfall from 600mm on the coast to 800mm in Namaacha"*.

However, according to FRANCISCO & PASSELA (2007:15) "the *climate of the study area is predominantly subtropical with an anti-cyclonic weather regime and mid-latitude depressions, with maximum rainfall in the hot season"*.

However, according to the National Meteorological Institute INAM (2010)[11] the average annual temperature recorded in Mafuiane (2009) was 23°C, with an average annual rainfall of 29.2mm. This demonstrates the change in the region's climate, since according to FRANCISCO & PASSELA (2007:15) quoting the IIAM (Umbilúzi station s/d) they state that:

> *"the average monthly temperature from 1952 to 1981 was 22° C with fluctuations from 18 to 26° C, the highest being in January and February (26° C) and the lowest 18° C between June and July. "The winds in the Mafuiane region are predominantly northeasterly (38.6% of the time) and easterly (31.0% of the time), with periods of predominance but without precipitation, although humid, with around 75% relative humidity."*

Looking at the graph below, we can see that the hottest month in 2009 was January, with a maximum temperature of around 27.45°C, and the coldest month was July, with a temperature of around 18.9°C. The annual rainfall (2009) was 351.2mm and January was the wettest month with around 129.9mm of rain, while the months from April to October were considered dry (see Fig.4 and Table 2).

For the local community, this behavior of the precipitation only harms their agricultural activity, as they have to rely solely on river water, which is already contaminated as a result of the intense mining activity in Mafuiane.

Table 3: Precipitation and temperature values (2009).

Months	Jan.	Feb.	Mar.	Apr.	Ma	Jun.	Jul.	Ag.	Sep.	Oct.	Nov.	Dec.
Precipitation (mm)	129.9	72.4	76.3	5.4	1.7	0.4	1.3	0	9.6	0	45.7	8.5
Temperature (°C)	27.4	26.5	25.3	23.6	21.9	19.7	18.9	19.8	22.2	23.7	25.1	26.3

Source: author, based on data obtained from INAM (2010) - State of Namaacha

Fig.4: Thermopluviometric graph (2009).

---

[11] See Annex 1.

Source: author, based on data obtained from INAM (2010)

2.2.3 Geology

From a geological point of view, the study area is part of the (CVL)[12] which is part of the South Save basin. Describing the geology of this region, PINNA, et al., (1986:212) states that:

> *"The filling of this basin by sediments began in the Cretaceous and was accompanied by transgressions and regressions caused by climatic variations and vertical movements that resulted in the formation of a series of grabens with a north-south (N-S) orientation. The base of the Basin is characterized by basaltic lavas and Karroo rhyolites (see Table 3)".*

From the same perspective, MUCHANGOS (1999:159) states that :

> *"The study area falls within the so-called CVL, where plains that constitute plateaus of rocks and forms of the Karoo volcanic complex predominate. These forms of the Karoo volcanic complex are the result of fissural volcanism with lava emitted in tabular layers, alternately acidic (rhyolites) and basic (basalts), locally interspersed with tuffy pyroclasts and breccias."*

The study area is essentially made up of two lithological formations chronologically belonging to the Karst and Cretaceous periods, basalt and intrusive rhyolite respectively. There are also rhyolitic series and recent diabase veins, rhyolitic flows and alkaline effusive rocks, tuffs, Quaternary formations, fluvial sediments, terrages and sedimentary deposits.

These two types of rock form the Libombos chain, which is a bimodal volcanic sequence with a north-south direction, forming a monoclinal 10°-20° to the east. This belt, according to AFONSO, et al. (1998:65), *"with a length of 800km and a width of 20-25km stretches from Pafúri to the border with Swaziland and South Africa".*

---

[12] See Annex 2

According to AFONSO, et al., (1998:65) the lithological sequence of this belt from top to bottom is as follows:

1. Upper Movene basalts, with intercalations of rhyolites from the Pequenos Libombos;
2. Umbelúzi rhyolites and ignimbrites;
3. Impaputo basalts, with sandstone intercalations in the lower part (not outcropping);
4. Lower Movene basalts.

Table 4: Summary of the Geology of the Southern Save Basin

Period		Marine origin	Continental origin	Lithology
Quaternary				Alluvial fans, terraces, dejection cones and lacustrine limestones
Tertiary	Superior		Mazamba training	Conglomerates, coarse sands and ferruginous sandstones
Tertiary	Superior	Jofane's training		Glauconitic sandstone, marl and limestone
Tertiary	Bottom	Salamanga Formation		Limestone and sandstone
Cretaceous	Superior		Elephant Formation	Reddish conglomeratic sandstones and siltstones
Cretaceous	Superior	Grudja formation		Marls, clays, glauconitic sandstone and evaporitic clay
Cretaceous	Bottom		Sena Formation	Conglomeratic sandstone with pebbles from eruptive rocks
Cretaceous	Bottom	Maputo training		Glauconitic limestone and saline marls
Jurassic	Karroo		Movene	Basalts, rhyolites and quartzolatites
Jurassic	Karroo		Umbeluzi	Riolites, tuffs and tranqui-andesites

Source: adapted by the author based on PINNA (1987).

2.2.2 Geomorphology

According to GUERREIRO & MUCHANGOS (2003), "*in the extreme west of the study area, near the border between Mozambique and Swaziland and/or South Africa, the rhyolites of the*

Grandes Libombos outcrop". The same authors state that this region *is:*

> *"made up of hills at relatively higher altitudes in the Libombos region, commonly known as the Great Libombos volcanic chain. This chain has a north-south orientation, the highest point of which is Mount Mponduine with an altitude of around 801m, the average for the region being 600m above sea level"* (Idem:2003).

To the east of the Great Libombos, there is a narrow plain made up of basaltic soils. This is bordered by a chain approximately 5km wide and around 50km long, the Pequenos Libombos chain. This chain has steep slopes to the west and gentle slopes to the east. However, according to FRANCISCO & PASSELA (2007:15) the study area has:

> *"a peneplanal morphology formed by Mount Secuane and the Umbilúzi depression with no accentuated terrain, rising slightly towards the west. The maximum heights reached by the relief in the quarry area do not exceed 96.85 meters in relation to the average sea level."*

The region is cross-cut by small north-south trending faults which are interpreted as being important in controlling the degree of alteration of the rhyolite and perlite that form the bentonite deposits that occur 5km from Mount Secuane towards Namaacha.

2.2.4 Soils

According to VILANCULOS & SERENO (1993:24) *"on the 1:5000 scale soil map, six (6) soil mapping units of Mafuiane were characterized* (see Table 4).

The different units mentioned above were classified based on the local geology and subdivided into the following characteristics: Soil color and effective soil depth.

Table 5: Soil types in Mafuiane and their characteristics

Units	Features
B1	These are lithic basaltic soils with a depth of less than 5 cm
BP1	These are black basaltic soils deeper than 100 cm
BP2	These are black basaltic soils between 50 and 100 cm deep.
BV1	These are red basaltic soils deeper than 50 cm
BV2	These are red basaltic soils between 50 and 100 cm deep.
S1	They are lithic soils on pebbles

Source: VILANCULOS & SERENO (1993)

According to the FAO classification (undated), this type of soil corresponds to

1. **B1:** Eutric Leptosols;
2. **BP1 and BP2:** Vertic Luv;
3. **BV1:** Vertic, Haplic Luvisols;
4. **BV2:** Haplic Luvisols;
5. **S1:** Mollic Leptosols.

VILANCULOS & SERENO (1993:24) carried out a detailed soil survey in the locality of Mafuiane using a topographic map on a scale of 1:50,000. Two observations were made with a natural probe at a depth of 1.20m. From this survey it was concluded that:

> *"Red basaltic soils (Bv) predominate in Mafuiane, occupying around 60% of the cultivated area. These soils are subdivided into red basaltic soils with a depth of more than 50cm. This region has soils with a loamy to clayey texture in the surface horizon and clayey in the subsoil with a subangular to angular anisoform structure."*

Therefore, they are shallow soils in certain areas due to the nature of the bedrock, as there are some medium-sized and fine loose stones scattered over the surface of the land, which are an obstacle to cultivation in the same areas.

It was also found that there are lithic basaltic soils on top of the pebbles, the depth of which does not exceed 50cm. Its consistency is hard to slightly hard when dry and friable *(fragmentable)* when wet, with little stickiness and little plasticity when wet. Drainage is moderate and low with basic infiltration of 50mm/h, moderate permeability.

Due to its location on one of the banks of the Umbelúzi River, Mafuiane has fertile soils rich in organic matter, suitable for farming, dark brown and black in color, with moderate to poor drainage, depending on the relief and geological conditions.

2.2.5 Hydrography

The mountain range where the Sulbrita quarry is located is not a watershed in a southerly direction. About 5km from the study area is the narrow valley of the Umbelúzi River. This area is drained by several watercourses that form part of the Umbelúzi Basin, most notably the Movene River, whose tributaries are: Chambadejovo, Boba, Maxibobo, Gumbe and the Impaputo.

The Movene and Impaputo rivers and their tributaries form a fairly dense network whose orientation follows the general slope of the relief and the morphological character of the region. The Movene River is the main watercourse in the study area. It and the Impaputo are one of the major tributaries of the Umbelúzi River, where the Pequenos Libombos Dam was built to supply water to the cities of Maputo and Matóla (see fig.5).

The Impaputo and Movene rivers have a periodic regime, with each of them having their main attributes at Namaacha District level: Chambadejovo, Boba, Maxibobo, Gumbe which are tributaries of Movene and the Impocuane, Mabelebele and Machuanine rivers which are from the Impaputo. Most of the tributaries and sub-tributaries on national territory originate in the Great Libombos.

There are also small streams that have developed over furrows dug into the rhyolitic formations, which generally have significant flows during the rainy season. With regard to groundwater, the study area is characterized by a very low water table.

Fig.5: Hydrographic map of the Namaacha Administrative Post

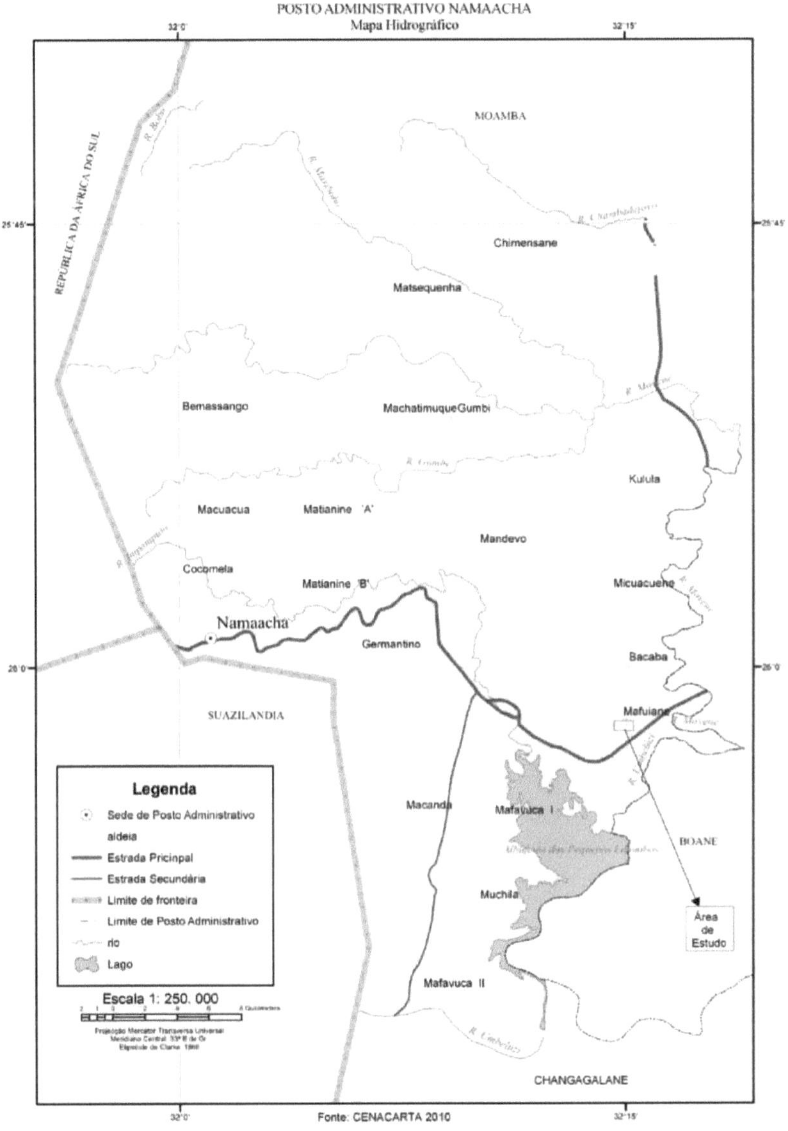

POSTO ADMINISTRATIVO NAMAACHA
Mapa Hidrográfico

Fonte: CENACARTA 2010

2.2.6 Flora and Fauna

The vegetation in the study area is part of the so-called Libombos Forest, which is distributed over areas of moderate altitude between 350 and 800m along the Grandes Libombos and Pequenos Libombos and is characterized by its normally deep, reddish, clayey rhyolitic soils.

It is a savannah-type vegetation characterized by isolated pastures and small trees integrated into the steppe. It has been and still is cut down by the population to obtain wood fuel, building material and charcoal for sale.

Basically, the Mafuiane locality is made up of emergent dwarf trees, shrubs between 0.5m and 3m high (see Fig.6).

However, MICOA (2004), quoting WILD & BARBOSA (1967) states that this type of vegetation "*indicates a great diversity of tree and shrub species. At present, the patches of regrowth forest are confined to the hard-to-reach areas of the Libombos mountain range*".

According to VILANCULOS & SERENO (1993:6) "*in the town of Mafuiane there are scattered Sclerocarya Caffra (canhoeiros), scattered shrubs and grass*". But grasses and xerophytic forest species are also found in abundance, as well as small forest galleries dominated by Adina Microcephala var Galpini, thorny mica trees stripped of their foliage during the dry season, Xilate, Mondzo and Nala (see Table 5).

In faunal terms, Mafuiane is dominated by small animals and a variety of reptiles and insects, while wild animals are practically non-existent.

Table 6: Predominant plant species in the study area.

Local name	Scientific name	Utility
Chivondzuana	*Combretum sp*	Wood fuel
Micaia	*Acacia sp*	Wood fuel
Canhoeiro	*Sclerocarya Caffra*	Traditional drink production
Mondzo	*Combretum imberbe*	Coal production
Xylate	*S. Africa*	Ditto
Nala	*Abizia sp*	For construction

Source: adapted by the author, based on VILANCULOS & SERENO (1993)

Fig.6: Land Use and Land Cover Map of the Namaacha Administrative Post

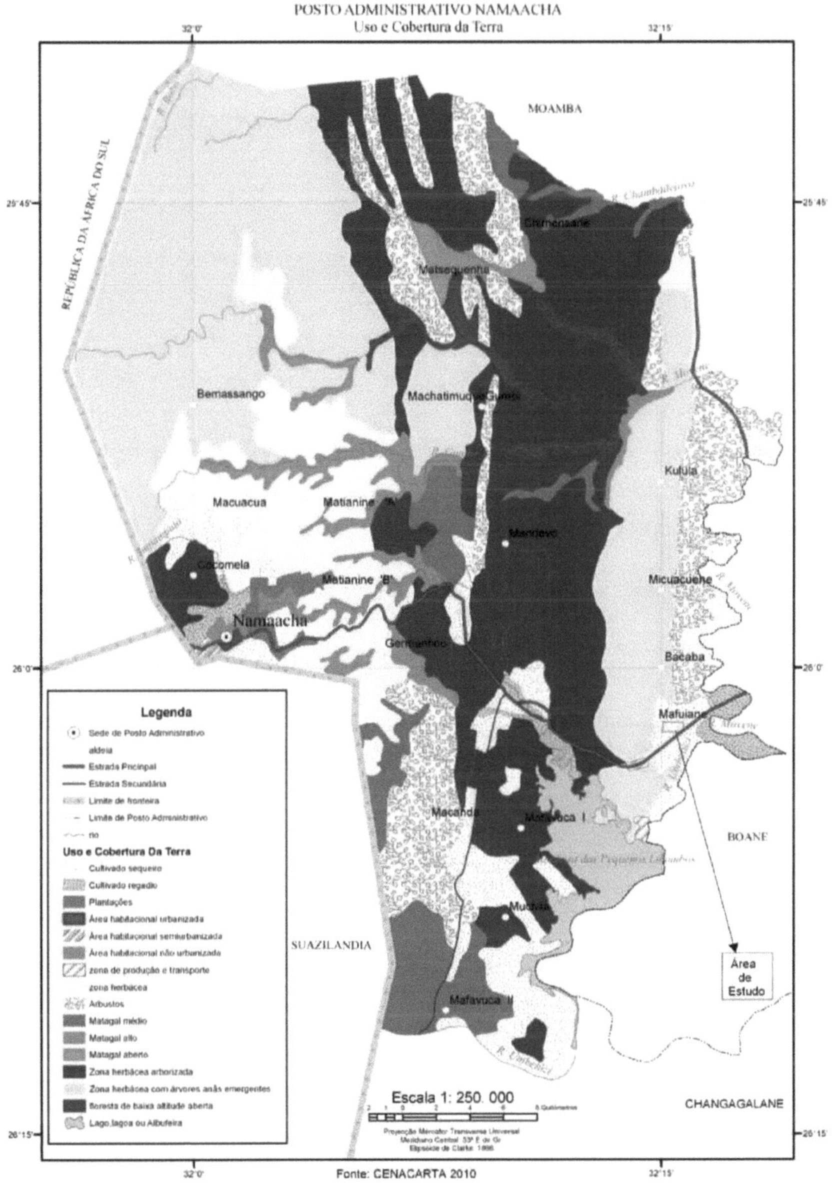

2.3 Socio-economic aspects

2.3.1 Population

According to the III General Population and Housing Census (III RGPH 2007), the town of Mafuiane has a total of around 2,648 inhabitants, which corresponds to 140 inhabitants per km^2 (Fig.7).

Fig.7: Graph of the evolution of Mafuiane's population (2007-2012)

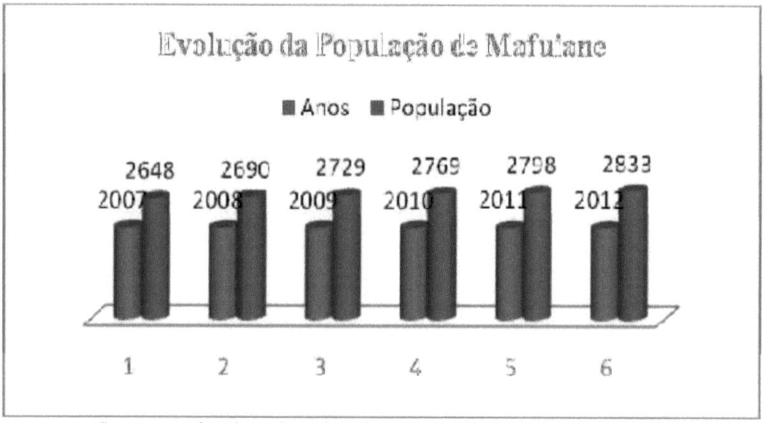

Source: author based on data from INE (2007)

The majority of the population living in the Bacabaca neighborhood is of Tsonga ethnicity, although there are some immigrants from the south of the country. The most widely spoken languages are Ronga and Xangana. The mainstay of these communities is agriculture and livestock farming. The entire socio-economic life of this region is concentrated in the towns of Boane and Namaacha, where it is possible to find some basic infrastructure.

2.3.2 Health

According to OMBE & FENHANE (2003:9) quoting the World Health Organization (WHO:2000)), health is not only the absence of disease, but the complete physical, mental, moral and social well-being of the individual.

The health network in the town of Mafuiane consists of a single conventional health center, whose current physical condition is considered reasonable, with no hospitalization, but with a maternity ward that serves the communities of Mafuiane, Makanda, Impaputo, Mafavuka and Micuacuene (see question 2 of Appendix 2). With regard to the problems faced by the Mafuiane Health Center, the following stand out:

> *"the lack of sufficient medical staff to assist patients, given that there is only one preventive medicine technician, a general medicine technician and a maternal and child health technician, including the lack of an ambulance to evacuate seriously ill patients who are kept under observation for 24 hours if necessary"* (MANGAZE, 2010, cp.)[14] .

The same source said that the center sees an average of 40 to 50 patients a day suffering from various illnesses. The most frequent cases are linked to diarrheal diseases, dysentery in adults, acute respiratory infections, bronchitis in children, malaria and HIV/AIDS (see questions 5 and 8 of Appendix 2).

The same source refers to the fact that several people are admitted to the health center every day with injuries, sometimes serious, caused by accidents in the stone extraction mines.

2.3.3 Polite

Education is a social phenomenon that can take many different forms and forms according to different social or human groups. It is also a process by which societies prepare their members to ensure continuity and development.

This sector, in the locality of Mafuiane, is organized as follows: Two (2) elementary school, namely Bacabaca Primary School (EP1) and Mafuiane Comprehensive Primary School (EPC), with a total of 843 pupils enrolled for the 2010 school year (Direcçao Distrital de Educaçao, 2010) (See table 6). Given the population projection for 2010, this figure represents 30% of Mafuiane's total population.

It should be noted that Bacabaca Primary School was built as part of the social responsibility

---

[14] See question 3 of Appendix 2.

of the company Pedreira da Sulbrita (see questions 7 and 22 of Appendices 1 and 4).

These students are assisted by a total of 20 teachers, nine (9) of whom are female. Of these, 16 teachers teach at EP1 level, one (1) of whom has no psycho-pedagogical training. Four (4) teachers at EP2 level, all with psycho-pedagogical training (Direcçao Distrital de Educaçao, 2010).

Table 7: Ratio of students enrolled for the 2010 academic year

Classes	Sex		
	Male	Female	Sub total
1 [a]	87	59	146
2 [a]	72	76	148
3 [a]	58	57	115
4 [a]	52	55	107
5 [a]	72	70	142
6 [a]	49	43	92
7 [a]	43	50	93
Total	433	410	843

Source: author, based on data from the District Education Directorate (2010)

2.3.4 Inhabited

The basic characteristics of the houses in the town of Mafuiane are of the palhota, mixed and basic type, which correspond to all the houses built from conventional plant material. They are called basic because they don't meet all the requirements to be considered conventional, i.e. they lack piped water inside the house, electricity, a kitchen and an internal bathroom. Thus, almost everywhere in the town, there are hut-type houses whose walls are covered in earth mixed with stone, with a mixed roof (see fig.8).

Fig.8 Photo of housing features

Source: author, 2010

2.3.5 Agriculture

The main activity in Mafuiane is agriculture. According to the Servidos Distritais de Actividades Económicas de Namaacha (SDAEN) in Mafuiane, rainfed and irrigated farming is practiced[15] by the family and private sectors respectively.

In the family sector, monoculture is practiced, characterized by the cultivation of maize, Nhemba beans, cassava, sweet potatoes and in some cases okra, tomatoes, maize, butter beans, cabbage, garlic, onions, peppers and cucumbers.

According to SDAEN (2010) *"the private sector practices irrigated agriculture. Polyculture is essentially practiced, with the predominant crops being bananas and sugar cane for the market."*

In terms of livestock, cattle and goats are raised in Mafuiane. According to the local community, these animals don't grow properly due to the scarcity of pasture, since it doesn't grow because of the dust that settles on the grass, preventing it from growing (see question 13 of Appendix 3 and fig.9).

Fig.9: Photo of the vegetation near the Sulbrita quarry (*M/c^j a/a thorny*)

Source: author, 2010

2.3.6 Transport and Communications

The town is crossed by the EN2, which connects Maputo City and Namaacha District. Communication within the locality is ensured by picadas, and the locality does not have any improved or tarmac roads.

"This sector is a major concern for the local authorities because in the rainy season the roads become impassable due to the nature of the local soil conditions. This situation makes it

---

difficult to transport agricultural products to market, especially during the rainy season. He also said that there is an activity plan to improve some roads, although it is not known when it will start" (NHANGUILA, 2010, cp.). [16]

There is no public transport service in Mafuiane. The population is taken to the town of Namaacha or to the cities of Maputo and Matola by a small number of the so-called "Chapa 100" or by stone transporters to Maputo City. *"There is only one semi-collective passenger transport linking Maputo City and Mafuiane. The other transport links the villages of Boane and Namaacha"* (Idem, 2010, cp.).

---

[16] See Question 2 of Appendix 1.

CHAPTER 3

THE ENVIRONMENTAL IMPACTS OF STONE QUARRYING IN MAFUIANE

3.1 Legal instruments for the exploitation of mineral resources in Mozambique

The relationship between the environment and the activities of the mining industry is regulated by specific legislation such as Decree 26/2004 of August 20, the Environmental Regulation for Mining Activities, the Basic Environmental Management Standards for Mining Activities approved by Ministerial Diploma 189/2006 especially dedicated to Level 1 mining activities, as well as general environmental legislation (the Framework Law on the Environment - 20/97).

This legislation directly relates the environmental aspects of mining[17] to the granting and continued use of the license to exploit minerals. The Mining Law (14/2002 of June 26) and the Environmental Regulation for Mining Activity classify mining activity into three levels, depending on the scale of the operations and the complexity of the equipment used, the documentary requirements for environmental compliance, and the costs involved.

Tier 1 mining activities are basically equivalent to the activities that the Regulation on the Environmental Impact Assessment Process classifies as Category C. Level 2 mining activities are broadly similar to Category B activities according to the Regulation on the Environmental Impact Assessment Process and Level 3 mining activities are broadly similar to Category A activities.

In other words, Level 1 mining activities must comply with basic environmental management standards; Level 2 mining activities with the Mining Law (Law 14/2002, of June 26), Article 37, number 1 and the Environmental Regulation for Mining Activities (Decree 26/2004, of August 20), Article 1, numbers 1, 2 and 3.

The regulation thus stipulates that Level 2 activities (quarrying, excavation of mineral resources for construction, mechanized prospecting and research and pilot projects), although the Environmental Regulation for Mining Activities does not stipulate that Level 1 and Level 2 activities must apply for a permit, it is to be assumed that all activities that could result in some environmental damage must undergo a pre-assessment.

However, in the case of mining, the Environmental Regulation for Mining Activities requires the applicant for an environmental license to submit a "program for the rehabilitation of the affected area and the closure of the mine" as part of the environmental management plan. To ensure that the closure of the mine will be carried out in accordance with environmental requirements, for Level 2

---

[17] Mining consists of operations and work related to prospecting, research, extraction, processing and treatment of mineral resources.

activities.

It is in this context that the extraction of any mineral in Mozambique requires obtaining the respective mining permit, and the Ministry of Mineral Resources is responsible for issuing reconnaissance, prospecting and research licenses, mining certificates and mining concessions.

Mining activity is covered by the Mogambican regulations on the Environmental Impact Assessment (EIA) process[18] , Decree number (n°) 76/98 of December 29, 1998, repealed by Decree number 45/2004 of September 29, 2004 issued by MICOA. This activity is also regulated by the Mining Law, Decree No. 14/2002 of June 26 and the respective regulation, Decree 27/2003 in conjunction with the provisions of Article 8 of the Environmental Regulation for mining activity, Decree No. 26/2004 of June 30 issued by the Ministry of Mineral Resources.

The other laws and regulations applicable to this activity are the Law on Land Tenure and Use, Decree 19/97 of October 1 on the use and exploitation of land and the respective Forest and Wildlife Law, Decree 10/99 of July 7 issued by the now defunct Ministry of Agriculture and Rural Development.

The Environment Framework Law 20/97 of October 1, 1997 applies to all public or private activities that directly or indirectly influence environmental components.
It is based in particular on the precautionary principle, which focuses on avoiding the occurrence of significant or irreversible negative impacts, regardless of whether there is scientific certainty that such impacts will occur.

The Environmental Framework Law stipulates that any activity likely to have a significant impact on the environment must be subject to environmental licensing, which is the result of an environmental impact assessment of the proposed activity and must precede any other licenses legally required in each case and the assessment of environmental impacts is based on an Environmental Impact Assessment (EIA)[19] to be carried out by entities accredited by MICOA.

The Mining Law, Decree No. 14/2002 of June 26, and its respective regulation, Decree No. 27/2003, regulate the terms of the exercise of rights and duties relating to the use and exploitation of natural resources with respect for the environment, with a view to their rational use for the benefit of the national economy.

---

[18]

According to SERRA & *CUNHA (2003:121), EIA "was officially enshrined for the first time in the USA through the drafting of the National Environmental Protection Act (NEPA), which came into force on January 1, 1970, and has been progressively adopted throughout the world, initially as a new legal technique, and currently as an authentic legal and environmental principle. It basically consists of the preventive submission of activity projects likely to cause more or less harmful effects on the environment to a process of investigation and analysis of a technical-scientific nature of those effects".*

[19]According to SERRA & *CUNHA (2003:125), an* Environmental Impact Assessment *(EIA) "consists of a document containing the necessary information on a given activity in order to inform the general public and support the decision of the competent authority on the environmental viability of that activity".*

Thus, in the study area, Monte Secuane is exploited by the Sulbrita Quarry, whose mining title was granted by the Ministry of Mineral Resources under No. 28/C/96, registered in the new mining cadastre system as Comissao Mineira 16C[20] and the respective regulations, with emphasis on liability for losses and damages, which state that:

> *"the holder of a mining concession who, by virtue of exercising mining rights, causes damage to crops, soil, buildings and improvements in the areas subject to the respective title, or causes the transfer of users or occupants of the land from the respective area of occupation, incurs the obligation to compensate the holder of the respective property and the resettled persons".*

Thus, the mining law requires mining operators to commission environmental assessment studies for level 2 mining activities, characterized by mining operations involving mechanized exploration methods and equipment.

3.2 Causes of Stone Extraction

The District of Namaacha is made up of a variety of CVL rocks from which stones, rhyolites, sands, clay and lime are extracted. Among the reasons for extracting stone in Mafuiane, we highlight the economic and social reasons, since mining contributes to local development and the economy of the district and the country. By way of example, according to the local chief and some residents, the construction of the Bacabaca Primary School by the Sulbrita quarry is mentioned, as well as supplying the material used in the rehabilitation of some access roads in the town and district if the local authorities request support in this regard (See Questions 4 and 9 of Appendices 1 and 2).

As we mentioned earlier, Mafuiane is home to various types of activities linked to the exploitation of mineral resources[21] (Fig.10).

Fig.10: Diagram of the main mineral resources exploited in Mafuiane

---

[20] See Annex 3.
[21] Mineral resources are any solid, liquid or gaseous substance formed in the Earth's crust by geological phenomena linked to it.

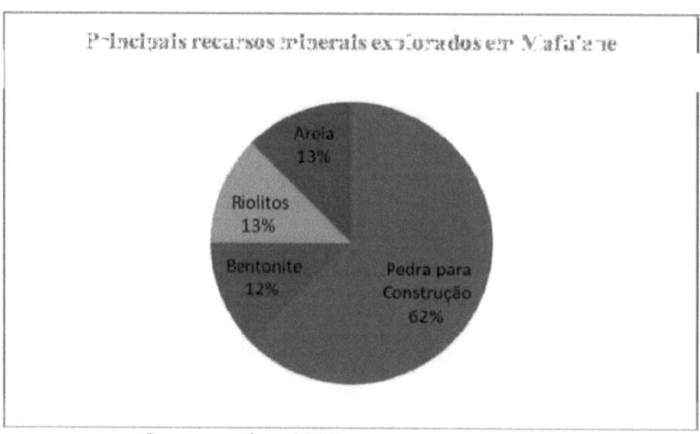

Source: author, 2010

The rapid rates of urbanization that the province and city of Maputo have experienced have created conditions for greater demand for this mineral for civil construction. This activity guarantees the subsistence and well-being of a large part of Mogambicans who continue to depend on natural resources directly linked to the extraction of stone. Continued socio-economic expansion, reflected in accelerated urbanization resulting from the development of the industrial and service sectors, population growth, among others, has increased the demand for natural resources and, in particular, for mineral resources.

3.3 Characteristics of the extracted rock

FRANCISCO & PASSELA (2007:9), characterizing the rock extracted by the Sulbrita quarry, state that:

> *"The rock extracted is a ridge of rhyolitic stone that protrudes from the basaltic substrate (Movene Basalt) forming a morphological unit called Secuane Hill. In general, the deposit shows an apparent homogeneity of the stone throughout the surface layer and depth, but it appears variably contaminated by clay stains and fractures that reach more than 5 meters deep. Beyond 5 meters, the rock is homogeneous and of high quality".*

It is an acidic rhyolite-spherolite rock from the granite family, with the same mineralogical composition whose gravel is reddish purple with greyish or reddish brown spots.

According to FRANCISCO & PASSELA (2007:9) the results of laboratory tests carried out by the Mogambique Engineering Laboratory (LEM) under registration number 23.643, *"the stone exploited on Mount Secuane can be used fundamentally for civil construction, roads (sidewalks, bases and sub-bases and structures) and the construction of buildings and pre-fabricated concrete elements".*

The same source states that *"the economic reserves of rhyolite on Mount Secuane have been determined at around 9,000,000m³ from the 1:1000 map, using the method of calculating the volumetric mass with commercial value"* (Idem., 2007:9).

3.4 Stone extraction process

The stone is extracted in the open air on benches, using mechanized equipment with the aid of explosives for the primary dismantling of the rock. *"The extraction process comprises a sequence of phases consisting of the removal of vegetation and soil, followed by the drilling of the rock using a rockdiller with an air compressor attached"* (NAMBURETE, 2010, cp.).

> *"The explosive charge is filled manually and blasted using a detonating cord and fuse retardant, followed by the disintegration of the rock using explosives. This is followed by transportation to the processing plant using wheeled shovels and transportation between the blasting front and the processing plant using dumpers"* (Idem., 2010, cp.).

"The material is deposited in the primary crushing section. From the primary crushing the material is transported by conveyor belts to a series of double vibrating screens (secondary screens) which separate the material into different fractions." (Idem, 2010, cp.).

According to the same source *"the quarry produces six (6) different fractions of material, namely stone dust (0-6mm), sarrisca (6-14mm), concrete (24-38mm) and rock (> 76mm)* (Idem., 2010,cp.).

3.5 Main Environmental Impacts of Stone Extraction in Mafuiane

Every cause has its corresponding effect, every benefit that man extracts from nature certainly has its downsides, in other words every human action in the natural environment causes some impact at different levels, generating alterations with varying degrees of aggression, sometimes leading environmental conditions to an even irreversible process.

The environmental impacts of stone quarrying in Mafuiane were identified on the basis of a qualitative analysis, using on-site visits and direct observation. A survey was also carried out of activities that could possibly cause environmental impacts, with the help of existing documentation on the quarries.

It can be seen that Monte Secuane has been exploited since 1992, which leads to the conclusion that although there are no actual documents proving the existence of environmental impacts in the region, judging by the life of the quarry and the type of activity carried out, environmental transformations take place on a daily basis. It is also worth remembering that the exploitation license granted to Sulbrita in 1996 is valid until 2021, i.e. 25 years and an estimated production of 20,000m^3 per month.

Through the observations and interviews carried out on site, it was possible to see that the region is facing one of the biggest and most serious problems of our time, which is environmental pollution caused by open-cast stone quarrying. It should be noted that we do not intend to address the positive impacts of the activity, but rather the negative ones, since they are the ones that cause the most concern in the community.

In addition to the degradation of the landscape, the main disturbances caused by stone extraction are vibrations, noise and suspended dust and gases from the operation of machinery, trucks and rock crushing plants, as well as disturbances caused to surface water or the water table.

The changes inherent in the activity also affect the fauna and flora, which are normally elements that can be restored or recovered, albeit partially. For a better understanding, the impacts have been divided according to the following stages or phases: infrastructure implementation and the operations phase (stone extraction).

3.5.1 Infrastructure implementation phase

At this stage, the impacts are smaller in scale and the effects are local, and can be grouped into: environmental and socio-economic, although they are also felt more severely during operations. At this stage, the vegetation cover is removed, favoring the compaction of the soil in the area destined for the installation of the quarrying infrastructures, the opening of access roads for the circulation of the machinery used in the quarrying of the stone (see fig. 11), and the fauna in turn is disturbed due to the destruction of its habitat[22] .

Also at this stage, as well as during operations, aquatic ecosystems are affected by various suspended solid substances, directly affecting terrestrial ecosystems as they depend on water.

As the activities pose a great risk to the population, they are forced to move to other regions that are theoretically safe, although this is not the case as the impacts are local and regional and sometimes long-lasting (see Appendix 5).

As the population moves to new areas, another problem arises: pressure on natural resources, which creates ecological imbalance. Not only does the population need a place to live, but the fauna is also forced to migrate to safer places where it can survive. As a result, other animal species that are less resistant to change end up becoming extinct.

Fig.11: Photo of some of the machines used to extract the stone

---

[22] Habitat according to GARRIDO and COSTA (1996:84) *"corresponds to the specific environment in which plants and animals are able to live, feed and reproduce"*.

Source: author, 2010

3.5.2 Operations or production phase

The most significant impacts at this stage include alteration of the natural landscape, contamination of surface and groundwater, air through dust and fumes, soil, vibration and noise.

3.5.2.1 Alteration of the natural landscape

The extraction of stone always has the effect of changing the landform on a local scale and over a long period of time, as vegetation, rocks and soil are removed in order to make the area more accessible for all quarry work.

The impacts caused by stone quarrying and its effects are evident and visible to the naked eye, as they in themselves cause a series of significant problems of both a landscape and pedological nature, given that the stone quarrying activities take place on the summit of the mountain and in the open air, so excavations are visible from long distances, as well as large masses of blocks piled up around the quarried area, directly affecting the visual aspect of the region's landscape (see fig.12).

Answering the question about the environmental transformations that occur with the extraction of stone in Mafuiane, Mrs. (AA)[22] and others said that there are several changes in Mafuiane that also affect the fauna and flora, pollution of the air, water, soil and a lot of noise every day. "*The mountain is ending and we're going to suffer a lot because of it. In this area we didn't suffer much from illnesses, but today we always get sick with coughs and diarrhea because of the dust that comes out of the quarry*" (See question 7 of Appendix 3).

The power of the machines used easily destroys large swathes of the exploited area in just a few years. Hills and other landforms are razed to the ground, transformed and the natural landscape replaced by waste rock. Large craters also appear, causing negative local impacts on the environment.

Fig.12: Satellite image illustrating changes to the natural landscape

Source: Google Earth provided on 05/20/2010

[22] Cfr. Table 8: Appendix 6 of the interviewees' code

3.5.2.2 Water contamination

Water is a natural resource that is indispensable to the life of living beings, but the mining process threatens its quantity and quality, as various chemical elements come into contact with the water, thus altering its quality. According to Mining Law No. 14/2002 of June 26, article 15 (water conservation), *"the right to use water for any mining activity will be exercised under the terms of the special regime established in Law No. 16/96 of August 3 and its regulations"*.

However, these regulations are not observed during the stone quarrying process, as the quality of the water in the rivers that flow near the quarrying area, and even in the boreholes that supply the local population, is not the best, as the rainfall and the force of the wind carry the material sedimented during stone quarrying into the watercourses (see fig. 13).

This situation is more critical during the rainy season[23] because all the waste rock and lubricants used in the quarry are washed into the rivers by the rain. It should be noted that we did not use any laboratory analysis to reach this conclusion. The area explored is crossed by the Movene and Uachanganine rivers, tributaries of the Umbilúzi river, which are the main sources of water for the community, used for various purposes including irrigation of agricultural fields.

From the observations made in the field, together with the data from the interviews and geological and environmental studies of the region, it was possible to see that the Movene River and

---

[23] See the Thermopluviometric graph.

other watercourses are silting up[24] with negative impacts on the aquatic ecosystem.

Suspended particles (silt and clay), including vegetation removed when the rock is cleared, are carried by rainwater and wind and deposited in watercourses. It is also contaminated by fuels, oils and greases used in machinery and trucks which, according to the communities, make the water unfit for consumption (see question 14 in Appendix 3).

It is fair to say that the recommendations of the Simplified Environmental Study Report are probably not being complied with, as it was expected that the water would be reused in the processing of the stone in a closed cycle, but the observations prove that it is not, as it is channeled through a ditch and deposited somewhere on the left bank of the EN2. The company has no wastewater treatment plant, which means that the water is not recycled. All the fuel spilled during the process of cleaning the machines joins with the water and flows into the streams.

Fig.13: Photo illustrating the appearance of the water in a section of the Movene River

Source: author, 2010

However, when contaminated, water is one of the biggest vectors for the transmission of diseases, allowing harmful elements to reproduce and when consumed it can cause various illnesses. The questions about the origin of the water consumed by the communities served to measure the degree of concern of the communities (What is the source of the water you use for consumption?). Responding to these questions, the interviewees (*AB, AC, AD*) and others replied that:

> *"We don't consume water from the local rivers as we did before the massive extraction of stone by the companies there, because according to them there are frequent cases of diarrhea caused by drinking the water. The Sulbrita quarry drains all the water used in the quarry into the rivers and the water always turns red and we are afraid to drink it even if it is boiling, as we did before. Even the water that the company offers us is full of residue and sometimes cloudy"* (See question 14 of Appendix 3).

---

[24]Siltation is the accumulation of debris (pebbles, sand, silt, etc.) in areas of low riverbed gradient.

The existence of diarrheal diseases in Mafuiane is supported by the local health workers when asked about the main diseases affecting the local population. Diarrheal diseases were mentioned first after coughs, which it is assumed are probably caused by the dust that the population located near the quarry in particular always inhales.

Water pollution not only creates problems for the health of the population, but also for agricultural activity. The town of Mafuiane is, by excellence, an agricultural area organized into two sectors: family and business.

The gentlemen (*AH, AI, AJ and AK*) were unanimous when answering the question about the development of crops in the fields, stating that:

> *"The quality of the water contributes to low agricultural yields in the family sector because crops do not grow satisfactorily when they are irrigated by river water due to their condition. We peasants here in Bacabaca suffer a lot from this quarry" (Idem.,2010.cp.).*

Our crops don't grow because the water we use to water them isn't right. When we water, the plants become small because of the water and dust. The leaves of the beans and other crops don't grow" (See question 12 in Appendix 3). Since watercourses are not stationary, the level of pollution in them ends up being regional[25] .

3.5.2.3 Air contamination (dust and fumes)

One of the vital necessities for human beings is air. It acts either by evolving man or by acting as a link between man and man. Air carries both inanimate and non-inanimate substances. Inanimate substances include dust, fumes and vapors resulting from human activities.

The air quality in the study area is directly affected by the suspended dust caused by the machinery used for cleaning, drilling, fragmenting the rock using explosives (dynamites), the movement of trucks between the extraction site and the crushing facilities and the storage of the stone until it is sold on unpaved roads, raising large quantities of dust and fumes into the atmosphere.

From this process, large quantities of dust particles are released into the atmosphere by the action of the wind, contributing negatively to local air quality over the medium to long term. Like the others, this impact is noticeable to the naked eye when the machines are in operation. Since the local relief is not homogeneous, the sediments removed at the top of the mountain are carried by rainwater and wind to the lower areas, including the watercourses, which are used by the population to irrigate

---

[25] See Table 9 (Appendix 5).

the fields.

This phenomenon confirms once again that the recommendations of the environmental impact study have not been complied with, as the report states that the potential sources of dust generation should be watered whenever necessary to prevent the dust from rising into the atmosphere. When asked about the consequences of the dust for the population, almost 95% of respondents said that:

> *"Dust is responsible for the poor quality of pasture and crops around quarries because when it lands on plants and soil it compacts to form a laterite armor, thus preventing the circulation of water and oxygen. The dust that comes out of the quarry causes various illnesses. Many people suffer from coughs and other illnesses that are foreign to us. Our children and the elderly are the most vulnerable"* (see question 7 in Appendix 3).

It is from this perspective that local communities complain of not being successful in their fields, as dust and other solid waste is deposited on the cultivated areas and on the crops, making it difficult for the crops to grow, as a laterite crust forms on the soil.

This situation becomes more aggravating during the rainy season because, given the characteristics of the soils, which are predominantly clayey and shallow, the deposition of dust makes it increasingly difficult for nutrients and water to circulate, thus favoring the accumulation of water on the surface, which contributes to low agricultural productivity. This situation is more prevalent in the areas adjacent to the quarry, specifically in the Bacabaca neighborhood and on the banks of the Movene River.

3.5.2.4 Vibration and noise[26]

We didn't use any measuring instruments to measure the acoustic parameters that characterize the noise in the study area, or to measure the degree of intensity of the vibrations, but through natural hearing and surveys we concluded that there are noise disturbances in the town of Mafuiane and that they are associated with the noise produced during all the work in the quarry, not just Sulbrita, but all the quarries that are working in Mafuiane, specifically the crushing machines and extraction equipment, as well as the vehicles that transport the processed stone to the consumer markets.

Originating mainly during the fragmentation of the rock, by explosions using dynamites and other chemical compounds, the vibrations spread throughout the rock mass, shaking neighboring buildings and causing physical damage to people and other living beings, as well as various assets, although there is no infrastructure worth mentioning, the communities living in Mafuiane say that when impulses occur, the aftershocks are felt everywhere.

---

[26] According to MELO & COSTA (1992:1723), vibration "is the act or effect of vibrating, oscillating or shaking. It corresponds to the rapid movement of the molecules of a sounding body when the complete shuttle of a vibrating particle is perished".
Noise: an inharmonious sound caused by a falling or cracking body, a bang, a roar. (idem:1471)

The survey data shows that 50% of the respondents living in Bacabaca regret the fact that their buildings are the daily targets of cracks and broken window panes resulting from explosions and rocks thrown from great distances, which hit houses and sometimes people, leading to constant deaths, especially in the vicinity of the Sulbrita quarry (see question 7 in Appendix 3).

This noise directly affects the quarry workers, causing them hearing problems, according to the interviewee (*NA*) and others, but also the population located near the quarry.

The results of the surveys carried out among the population of Mafuiane indicate that around 99% of those questioned say that there are loud explosions in the quarries every day at around 5pm. And when this happens, the population is always forced to take refuge in the bush in order to protect themselves from the stones that are thrown from great distances, because according to our findings, people have died in their homes when they are hit by stones (see question 10 of Appendix 3).

3.6 Mitigation measures

Once the stone has been extracted and it has been found that the area no longer offers conditions for the process to continue, rubble should be used to cover the exploited and abandoned areas, as well as replanting the vegetation, because this is the only way to minimize the visual and landscape impact. At the moment, however, this is not the case, as almost everything extracted from the quarry is destined for the consumer market, and there is no rubble to cover the open craters.

Conditions should be created for the retention of all water used in the extraction of stone so that it does not reach watercourses before being recycled, thus reducing the content of waste and lubricants in the stone. This is unfortunately not the case at the Sulbrita quarry.

Dust can be reduced by adopting specific measures, such as paving the main internal roads with an asphalt layer or humidifying the roads frequently. The company should abandon the dry extraction method and use water to constantly humidify all crushing sections to avoid emitting greater quantities of dust into the atmosphere.

In addition to this method, of course, dust can be reduced by intensifying the installation of suitably arranged vegetation curtains or by promoting the regeneration of natural vegetation.

It should be noted that this practice has been implemented by the Sulbrita quarry, but given the intensity of production, we believe that it is not enough to combat the enormous amount of dust in the area. These measures help not only to limit the effects of air displacement, but also to contain dust and noise.

Vibration and noise pollution can be reduced by using new operating technologies, for example the installation of silencers and constant maintenance of the machinery used to ensure that noise emissions are within acceptable levels. All workers should have noise protection equipment, something that is currently not the case, as only one and one other have earphones to protect against hearing damage, specifically those who are directly connected to the crusher in the primary section.

Since the vibrations directly affect the communities closest to the quarries, it would be advisable for the quarry to create mechanisms to remove the population from the vicinity of the quarry, so as to avoid the aftershocks of the explosions and the stones thrown up during the explosions frequently hitting the homes and residents of Bacabaca, creating damage that is sometimes irreparable, such as the death of the population.

4. Approach to the theme in Geography Teaching (1st and 2nd Cycles)

Geography has a broad scope in school curricula. As a result, it is possible to build values for group life, developing awareness of the existence of diverse peoples with different cultures, where these differences need to be respected and, at the same time, the interaction of these people with the environment in which they live is observed.

Therefore, aspects related to the use, conservation and protection of nature are included in the Geography teaching programs at the level of General Secondary Education (1st and 2nd Cycles) (see Table 8).

Table 9: Insertion of the theme in Geography teaching programs

Class	Objectives	Contents
8ª	Assume a critical, active and conscious attitude towards the use, conservation and protection of natural resources.	Use, protection and conservation of the atmosphere and biosphere; Use, protection and conservation of the Hydrosphere; Use, protection and conservation of the lithosphere.
9ª	Apply geographic knowledge in the community to prevent and mitigate different natural, social, economic and other problems.	The use of natural resources and environmental problems; Sustainable development; Protection and conservation of natural resources.
10a	Analyze the impact of industry on the environment	Extractive industry in Mozambique
11a	Take a critical attitude towards the importance of protecting and conserving bioclimatic resources.	The importance of protecting and conserving bioclimatic resources.

12a	Analyze the impact of industry on the environment; Relating industrial development to the use of natural resources; Explain the impact of industrial activity on the environment.	The impact of industrial activity on the environment. Protection and conservation of natural resources (Sustainable Development).

Source: author based on data from MEC (2008)

In order to achieve these objectives, geography teachers need to be more active in disseminating methods of using, conserving and protecting natural resources to ensure sustainable development.

FINAL CONSIDERATIONS

Conclusions

Aware that science is an unfinished journey, and that the conclusions of any research should be situated at the time they are formulated, the literature review and the study carried out have allowed us to draw some conclusions, contributing to a clarifying synthesis of the work carried out both in the office and in the field.

The subject of this study was the environmental impacts caused by the Sulbrita quarry in the town of Mafuiane. To this end, several descriptive analyses were carried out, based on theoretical assumptions combined with practice, analyzing the physical-geographical and socio-economic aspects of the town of Mafuiane, with a view to understanding the dynamics of mining in the region, given that not only stone is extracted for construction, but also various minerals useful to society, such as bentonite, rhyolites and others.

Specifically, the aim was to assess the impacts that this activity has on the environment, which became clear during the course of the work that, in addition to guaranteeing the economic sustainability of some of the families that depend on it, in particular those that are directly linked to the extraction of stone, i.e. those affected by the Sulbrita quarry.

This activity causes various types of damage to the environment, specifically water pollution, soil pollution, air pollution, noise pollution, vibrations and alteration of the natural landscape, also known as visual or aesthetic impact, which leads us to conclude that this is due to non-compliance with the rules established in the mining licenses granted by the competent authorities, including the recommendations of the Final Report of the Simplified Environmental Study carried out in 2007.

From this perspective, it can be concluded that the Sulbrita quarry has done little to reduce environmental damage, and the activity is taking place without complying with the established rules, since the only measure taken by the company is to plant eucalyptus trees that serve as windbreaks, thus reducing the amount of dust that can reach the cultivated areas, but this is not enough given the intensity of the extraction of stone at the site.

As a result, a number of problems have arisen that are reflected in the social life of local communities, particularly those that are located and carry out their economic activities in the area adjacent to the quarry, as well as jeopardizing the public health of the community in general and the workers themselves.

Recommendations

Given the findings made on the ground, we recommend the following: That the quarry be more attentive to environmental issues, specifically its conservation and preservation in accordance with the regulations established by the mining law, fully complying with the regulations contained in the mining license granted by the regulatory authority for this activity.

Mechanisms must be put in place to relocate the population near the quarry, as the damage not only affects the environment, but also human beings, given the constant deaths that the communities complain about and the damage to their homes.

The recommendations extend to local authorities and beyond to be more vigilant about complying with established regulations, because only then will the damage be minimized.

That the penalty rules laid down by law be applied in the event of non-compliance with established standards, not just for the company in question but for all those carrying out this activity in this region.

BIBLIOGRAPHY

AFONSO, Rui S. et al. *A evoluqáo da geología de Moqambique.* Maputo, Direccao Nacional de Geologia, 1998.

. *A Geología de Moqambique: Noticia Explanativa da Carta Geológica de Moqambique (1:2000.000).* 2.ed.Maputo, Direcgao Nacional de Geologia, 1978.

ALAMEIDA, António. *Environmental education: the importance of the ethical dimension.* Lisbon, Livros Horizonte, 2007.

BRAGA, Benedito et al. *Introduqáo a Engenharia ambiental: o desafio do desenvolvimento sustentável.* 2.ed. Sao Paulo, Pearson Prentice Hall, 2005.

BRANCO, Samuel Murgel. *The Environment in Debate.*11.ed. Sao Paulo, Editora Moderna, 2000.

BOLEO, José de Oliveira. *Physical Geography of Mozambique.* Lisbon, 1950.

COSTA,J.Almeida & MELO, A.Sampaio. *Dicionário da Lingua Portuguesa.*2.ed. Porto, Porto Editora, Lda, 1992.

FAO-UNESCO. Soil *Map of the World,* World Soil Resources Report, preliminary version in Portuguese, 1995.

FRANCISCO, Dulcídio & PASSELA, Paulo. *Final Report of the Simplified Environmental Study.* Maputo, 2007.

FELLENBERG, Günter. *Introduction to the Problems of Environmental Pollution.* Sao Paulo, Editora Pedagógica Universitária Ltda, 1980.

GARRIDO, Dulce & COSTA, Rui. *Brief Dictionary of Geography.* Lisbon, Presenca Publishing House, 1996.

GUALE, Rosaque Joao. *Potencialidades da Bacia do rio Movene para a construcao de Barragem,* Paper for the degree of Licenciatura (unpublished). Maputo. UEM/FL, Department of Geography. 1999.

GUERREIRO, F.M & MUCHANGOS, A dos, *Igneous rocks (volcanic intrusives) of the great Libombos (Field guide).* Maputo, 2003.

HATTON et al. *Evaluation of the Environmental Impact of Agricultural Practices around the Pequenos Libombos Dam.* Maputo, 1993.

INSTITUTO NACIONAL DE ESTATISTICA, *III Recensamento Geral de Populaçao e Habitaçao.* Maputo, 2007.

LAKATOS, Eva Maria e MARCONE, Marina de Andrade. *Metodologia científica.* 2.ed. rev ampli. Sao Paulo, Atlas, 1991.

MICHAEL at tal. *State of the Environment.* Lisbon, Instituto Piaget, 1993.

MINISTRY OF STATE ADMINISTRATION (MAE). *Profile of the district of Namaacha Maputo Province.* Maputo, MAE, 2005.

MINISTRY FOR THE COORDINATION OF ENVIRONMENTAL ACTION (MICOA), *Agenda 21 in Mozambique: National Council* for *Sustainable Development*. Maputo. MICOA, 2002 .
Environmental Strategy for the Sustainable Development of Mozambique. Maputo. MICOA.2004.

__________ . *National Environmental Management Program*. Maputo, MICOA.1995.

MINISTRY OF EDUCATION AND CULTURE (MEC). *Geografia: Programa da 8ª Classe*. Maputo, MEC, 2008.

__________ . *Geography: Grade 9 Program*ª . Maputo, MEC, 2008.

. *Geografia:* Programa da 10ª Classe. Maputo, MEC, 2008.

. *Geography:* Grade 1 program. Maputo, MEC, 2008.

__________ . *Geography:* 12th Grade Program. Maputo, MEC, 2008

MUCHANGOS, Aniceto dos. *Landscapes and Natural Regions*. Maputo, Tipografía globo, 1999

__________ . *Environmental Education: Fundamentals and Strategies*. s/ed. Maputo, DINAME, 2007. NHATUGÊS, C.L. *Petrography and geochemistry of the Great Libombos*. Degree work. UEM: Faculty of Letters, Department of Geography, 2003.

NOTICE, Joaquim. *The discourse on environmental education in 10th grade Geography in the context of the defense and conservation of nature in Mozambique*. Master's thesis in Education/Curriculum, Sao Paulo, Pontificia Universidade Católica de Sao Paulo in agreement with Universidade Pedagógica, 2006.

OMBE, A Zacarias & FENHENE, Joao. B. *Noçoes de Geografia Médica*. Maputo, MICOA, 2003.

PASSELA, Paulo. *Environmental Impact Study Report*. Maputo, Colégio e Consultores Ambiental, 2006.

PINNA, P, et al. *Popular Geological Map of Mozambique, scale 1: 1,000,000*. Maputo, National Institute of Geology (ING), 1987.

PIRES, Elizete Maria Pinto. *Monograph of the District of Namaacha*. Maputo. UEM: Faculty of Letters, Department of Geography, 1995.

SACATE, Jerónimo Ángelo. *The impact of drought in Kala-Kala Locality from 2002 to 2007*. Graduation paper (unpublished), Maputo, Pedagogical University, Department of Geography, 2009.

SALOMÁO, Alda. *Commented environmental law*. Maputo, Centro de formagao jurídica e Judiciaria, 2006.

SZABO, Atlla. *Research into building stone and Movene ballast*. Maputo, National Geology Institute (ING), 1985.

SERRA, Carlos Jr & CUNHA, Fernando. *Manual of Environmental Law*. Maputo, Centro de formagao jurídica e Judiciaria, 2003.

SERVIAOS DISTRITAIS DE ACTIVIDADES ECONÓMICAS DE NAMAACHA (SDAEN),

Annual Report. Maputo, Namaacha District Government, 2009.

NAMAACHA DISTRICT EDUCATION, YOUTH AND TECHNOLOGY SERVICE (SDEJTN). *Annual Report*. Maputo, Namaacha District Government. 2009.

PEDAGOGICAL UNIVERSITY. *Norms for the Preparation and Publication of Scientific Works at the Pedagogical University*. Maputo, UP, 2004.

VILANCULOS, M & SERNO, G. *Detailed soil survey of the Mafuiane-Namaacha area, Maputo Province*. INIA, 1993.

Republic of Mozambique. Bulletin of the Republic: Approves the Framework Law on the Environment. in: Bulletin of the Republic, October 1, 97.

__________ . Boletim da República: Approves the Mining Law. in: Boletim da República of June 26, 2002.

__________ . Boletim da República: Approves the Land Law in: Boletim da República of October 1, 97.

__________ . Boletim da República: Approves the Law on Forests and Wildlife. in: Boletim da República of July 7.

__________ . Boletim da República: Approves the Environmental Regulation for Activity

Mining. in: Boletim da República of August 20, 2004. . Bulletin of the Republic: Ministerial Diploma 189/2006-Basic Environmental Management Standards for Mining Activities.December 14, 2006.

__________ . Bulletin of the Republic: Decree 76/98 of December 29, 1998, repealed by Decree No. 45/2004 of -approves the process of Environmental Impact Assessment. in: Bulletin of the Republic of September 29, 2004.

INTERNET

United Nations Organization (UN). *International Year of Mountains*. Available online at: http://www.montanhasbrasil.org.br/ano.htm accessed on December 20, 2009.

Printed by Books on Demand GmbH, Norderstedt / Germany